Semirames Silva
Francisco Sousa
Eliezer Siqueira

Social characteristics and production techniques of family farmers

Semirames Silva
Francisco Sousa
Eliezer Siqueira

Social characteristics and production techniques of family farmers

ScienciaScripts

Cover image: www.ingimage.com

This book is a translation from the original published under ISBN 978-3-330-76187-2.

Publisher:
Sciencia Scripts
is a trademark of
Dodo Books Indian Ocean Ltd. and OmniScriptum S.R.L publishing group

120 High Road, East Finchley, London, N2 9ED, United Kingdom
Str. Armeneasca 28/1, office 1, Chisinau MD-2012, Republic of Moldova, Europe
Managing Directors: Ieva Konstantinova, Victoria Ursu
info@omniscriptum.com

Printed at: see last page
ISBN: 978-620-8-37950-6

SUMMARY

CHAPTER 1

INTRODUCTION

Primitive societies cultivated the land through traditional agricultural practices in order to obtain food for their survival. They maintained a balance between man and nature and preserved traditional cultures (rituals, symbologies, seeds, knowledge passed down from generation to generation). Over the years, man has perfected his work tools and created new technologies, leading to the appearance of machines and many agricultural implements (LOPES; LOPES, 2011).

Family farming is characterised by the diversity in the organisation of its internal structure, in terms of the availability of the use and distribution of resources - land, labour and capital (GERARDI; SALAMONI, 1994). This segment has become the focus of study, especially with regard to the strategies adopted to organise and reorganise itself in the face of the specificities of the capitalist mode of production.

Many works continue to be produced with the aim of deepening knowledge about family production in agriculture, speculating on its fate, the ways in which this segment will develop in the contemporary capitalist system of production, its process of adaptation to the market system, its development parallel to the capitalist system.

For Borges Filho (2005), the new scientific and technological bases of agricultural research are directly related to the concept of sustainable agriculture. In this sense, the current agricultural research framework is moving in the direction of developing more environmentally sustainable technologies, such as biological control, pest and disease monitoring, proper soil management, environmental impact assessment, among others.

The term sustainable agriculture refers to the search for durable, long-term yields through the use of ecologically appropriate management technologies, which requires optimising the system as a whole and not just the maximum yield of a specific product (ALTIERI; NICHOLLS, 2000). For Duarte (2012), sustainable development models are understood to be all production and/or processing systems that promote economic growth while guaranteeing respect for the environment, as well as the social and economic environment, with maximum integration and utilisation of available resources.

Faced with the dilemma between the growing need to produce food and the need to preserve the environment, ecologically-based agriculture has emerged as an alternative, with family farmers being the main players in building this bridge. Schroetter (2010) explains that family farming has come to be seen as a way of generating employment and productive occupations in the development of society.

However, according to Meirelles (2004), the high level of capital investment excludes a significant proportion of family farmers from access to technology, since the use of machinery and high-yield seeds is incompatible with the steep slopes and low fertility of family production units. These are the starting points for bringing family farming closer to ecologically-based agriculture, where the alternative has been to develop mechanisms for adapting to and coexisting with agro-ecosystems, in a process of observing nature and generating technologies that are compatible with the local reality.

In the 1990s, with the Eco-92 event, the concept of ecological agriculture broadened and brought a more integrated and sustainable vision between the areas of production and preservation, seeking to rescue the social value of agriculture, becoming known as Agroecology (FEIDEN, 2005).

Thus, the term sustainability, in its broadest sense, becomes of greater interest to agroecology. It is inseparable from rural development aimed at family farming, which requires a transdisciplinary approach, providing a cultured and fruitful dialogue between the natural sciences and the human and social sciences (IAMAMOTO, 2005).

In this sense, it can be said that "agroecology - ***the*** scientific and technological basis for a sustainable rural development project" is based on recognising the different rationalities of production decisions present in family production. This involves much more than organic farming, as its ultimate goal is to build a new concept of rural development (RIBEIRO; SALAMONI; COSTA, 2009).

In view of the above, the aim of this study was to characterise the social, environmental and economic aspects of family farmers in the floodplains of Sousa-PB. To this end, determined the following specific objectives: i) to distinguish the agroecological producers of the Sousa floodplains; ii) to identify the agroecological practices used by these producers, based on the principles of sustainability; iii) to verify the possibilities and restrictions for the development of sustainable family farming among these producers.

CHAPTER 2

LITERATURE REVIEW

2.1. Agribusiness and the Brazilian Economy

In Brazil, the concept of agribusiness emerged in the 1980s and was formerly recognised as the Agro-Industrial Complex. Over the years, agribusiness has evolved to involve the country's entire production chain, involving the purchase of seeds, seedlings, fertilisers, agrochemicals, tractors and implements, irrigation equipment and packaging. With the production of coffee, papaya, soya, corn, rice, beans, fruit, vegetables, planted forests, livestock, agrotourism, among others (OLIVEIRA, 2008).

According to Escóssia (2009), agribusiness can be conceptualised as "the sum of the operations of production, circulation and distribution of agricultural supplies, also encompassing the set of all agricultural and business operations", and should consider investments in research, as well as the production, processing and commercialisation sectors.

Some groups make up the trade balance of Brazilian agribusiness, such as chemical fertilisers and pesticides, animal products and the food industry. These groups have developed through the technological revolution, with increases in productivity that have contributed to the competitiveness of agribusiness in Brazil. Soon after the Real Plan, in the 1990s, there was growth in this sector, which led to an increase in production accompanied by falling prices for consumers (BARROS; GOLDENSTEIN, 1997).

According to Araújo et al. (2004), "Brazilian agribusiness has been gaining representativeness in national production year after year, especially in exports". The authors explain that according to IBGE data from 2003, agribusiness grew by 1.6 per cent, a rate higher than the overall growth of Brazil's GDP (0.3 per cent). This is due to the increase in industrial products used by agriculture, with the machinery and equipment segment also showing significant growth rates, particularly 24.4 per cent in 2003.

In 2009, agribusiness accounted for 27 per cent of the Gross Domestic Product (GDP), generated 37 per cent of jobs and accounted for more than 40 per cent of exports. It has become a sector of the Brazilian economy with a trade balance surplus. At that time, there were approximately 4.9 million rural establishments, equivalent to 60% of traditional agriculture, with low technological use, and family businesses were part of this model (ESCÓSSIA, 2009).

In 2013, Brazilian agribusiness is evolving with a focus on competitiveness and modernity, using technology in pursuit of sustainability. With the potential for 26 prosperous, safe and profitable production activities, it stands out for its production of grains, meat and commercial forestry plantations. The use of technology makes it possible to make better use of the soil, reducing the use of agrochemicals, as well as reducing the greenhouse gases that cause global warming (BRASIL, 2013).

Cotton production, for example, in the states of Mato Grosso, Bahia and Goiás, accounted for 88.7 per cent of the country's production in 2013, contributing to economic growth (BRASIL, 2013). It can also be seen that:

> Projections for cotton lint indicate production of 1.35 million tonnes in 2012/2013 and 2.53 million tonnes in

2022/2023. This expansion corresponds to a growth rate of 5.1 per cent per year during the projection period and a variation of 87.6 per cent in production. Consumption of this product in Brazil is expected to grow at an annual rate of less than 1.0
% over the next ten years, reaching a total of 915,000 tonnes consumed in 2022/2023. Exports are also expected to expand strongly, by 58.7 per cent between 2013 and 2023 (BRASIL, 2013).

Thus, according to the Ministry of Agriculture, Livestock and Supply (MAPA), projections indicate an increase in production of other grains, such as rice, beans and corn. Rice production is projected to increase by 11.1 per cent over the next 10 years. Beans are the product whose production is most in line with consumption, with an expected change in production between 2012/13 and 2022/23 of 14.2 per cent. Maize, on the other hand, is expected to produce between 78.8 and 89.0 million tonnes between 2013/14. And for 2022/23, projected production is 93.6 million tonnes. Other products that stand out in these projections include coffee, milk, sugar, meat, fruit, soya and tobacco, all of which make a significant contribution to the country's economy, especially with the export of some products (BRASIL, 2013).

Barros (2006) comments that Brazilian agribusiness has a variety of products that are structured into a production chain, including: sugar and alcohol, oranges, coffee, soya, cotton, wood (furniture, pulp and paper, plywood, etc.), tobacco, rubber, cocoa, fruit, tomatoes, red meat, chicken meat, pork, eggs, milk, potatoes and tomatoes. With lower productivity in the cultivation of flowers and vegetables.

Thus, this variation guarantees production stability, with natural price variations that have less effect on the agribusiness system as a whole (BARROS, 2006). Brazil has a large domestic consumer market as well as diversified exports. According to Duarte (2012) family farming alone produces the main foods consumed by the Brazilian population: 84 % of cassava, 67 % of beans, 54 % of milk, 49 % of corn, 40 % of poultry and eggs and 58 % of pigs. In the Northeast, family farming is responsible for 82.9% of the labour force in the countryside.

Conventional agriculture has been characterised since the colonial period by full-sun monocultures with a low level of biological diversity, disregarding the idea that plants can be grown in polycultures (AGUIAR-MENEZES et al., 2007). The practice of conventional agriculture, especially with the advent of the Green Revolution, developed in a way that did not take into account the damage to the environment (FRANCO, 2010).

The aim of the Green Revolution was to expand agricultural production so that it could keep pace with population growth (FRANCO, 2010). In this sense, conventional agriculture was based on the means necessary to increase production, through the use of machines and implements to prepare the soil, as well as the use of soil improvers and fertilisers, and the use of products to control pests and diseases when planting.

For Meirelles (2004), this model, based on the cultivation of highly productive genetic varieties, the use of chemical-synthetic techniques and inputs, mechanisation and the use of non-renewable energy sources, is responsible for the growing deterioration of agricultural systems.

The adoption of an agricultural production model based on technologies produced by the so-called "Green Revolution" makes us realise that conventional agriculture does not address the real socio-economic conditions faced by farmers and society. Since it is an unsustainable model, both socially and economically, alternatives to this form of production have been discussed all over the world.

In general, according to Lopes and Lopes (2011), the conventional agricultural system is characterised by the artificialisation and simplification of agro-ecosystems, generally made up of genetically similar or identical plants that have been selected with the aim of increasing productivity, and is highly dependent on external inputs such as soluble fertilisers, machinery and fuel. This causes an ecological imbalance, altering the processes of self-regulation of pests and diseases, which possibly reduces the recovery power of agro-ecosystems related to climatic adversities.

Modern agriculture has transformed fields into veritable production machines, replacing the artisanal production process based on the hoe, animal traction and, above all, natural fertilisers, with industrialised technologies based on chemical fertilisers, tractors, improved plant varieties and chemical pesticides (BORGES, 2000).

The modernisation of agriculture has followed capitalist lines, favouring what is known as the industrialisation of agriculture, making it an essentially entrepreneurial activity (TEIXEIRA, 2005). As a result, techniques, innovations, practices and policies have led to the degradation of natural resources (soil, water sources, genetic diversity, biodiversity), creating a dependency on non-renewable fossil fuels (GLIESSMAN, 2005).

According to Borges Filho (2005), the practice of agriculture by man involves simplifying the original ecosystem, favouring destabilising factors and forcing farmers to resort to energy-intensive techniques to maintain favourable conditions for plant growth.

Because they are very simplified ecological systems, monocultures are very unstable, favouring the establishment, multiplication and spread of pests, diseases and invasive weeds. In this way, simplified agroecosystems require frequent applications of pesticides (insecticides, fungicides, herbicides and others), causing other environmental problems (BORGES FILHO, 2005).

Worsening environmental impacts have contributed to increasing consumer demand for healthier agricultural products that are less harmful to the environment. These factors motivate more studies in more ecological lines of research.

2.2. Family Farming

Family farming has very diverse origins in the various regions of Brazil. The socio-economic and political invisibility of family farming was the result of a long process of subjugation and, in many cases, dependence on large-scale export agriculture. The large estate, dominant throughout Brazilian history, imposed itself as a socially recognised model (MOTTA; ZARTH, 2008).

Family farming includes all family-based agricultural activities and is linked to various areas of rural development. Family farming consists of a means of organising agricultural, forestry, fishing, pastoral and aquaculture production that is managed and operated by a family and predominantly relies on family labour, both women and men.

Family farming is made up of small and medium-sized producers and represents the majority of rural properties in the country, with around 4.5 million establishments (PORTUGAL, 2004). It is the mainstay of agricultural production for domestic supply, it produces basic foodstuffs that feed a large part of the population and it is the main source of income for many Brazilian families, playing an important role in the national economy.

According to Chonchol (2005), hunger is not only the result of insufficient food production, but also of the economic marginalisation of certain populations. Therefore, the issue does not revolve around increasing the production of those who already produce a lot, but rather giving everyone the means to produce. The discussion about the importance and role of family farming has been gaining momentum, driven by debates based on sustainable development as well as the generation of employment and income.

According to Wanderley (2003), family farming is a generic concept that incorporates multiple specific situations, with the peasantry being one of these particular forms. The transformations that have taken place in modern family farming cannot be seen as a total break with peasant forms, because it is these peasant characteristics that keep it strong, capable of adapting to the new demands of society.

Since the 1990s, there has been a growing interest in family farming in Brazil, through public policies such as the National Programme for Strengthening Family Farming (PRONAF) and the creation of the Ministry of Agrarian Development (MDA), as well as the reinvigoration of Agrarian Reform (OLALDE, 2004).

According to Mesquita and Mendes (2012), family farming is a concept used to characterise rural production units, structured around family labour, which are identified by the relationship between land, work and family. This mode of production has its origins linked to the history of the colonial regime and has always been related to the socio-economic transformations experienced in the countryside.

However, as Simon (2014) points out, despite their significant participation in the Brazilian economy, there is still a reality marked by poverty. "Almost half of the 24 million miserable people live in rural areas. For all these reasons, 2014 aims to reposition family farming at the centre of agricultural, environmental and social policies", in order to identify efficient ways of supporting farmers, and is dedicated as the International Year of Family Farming.

When analysing the Brazilian debate on family farming, Schneider and Plein (2003) state that in the political field, the adoption of the term seems to be related to the clashes that social movements had in discussions about space and the role of small rural producers. On the other hand, this adoption of the term came about through some academic works that began to look for new theoretical and analytical references in this period, which contributed to the introduction of the term family farming.

Since then, studies have pointed to the importance of family farming in the country's socio-economic context, as it is an activity that, in addition to generating employment, allows farming families to remain in the countryside, reducing the rural exodus and overpopulation in urban areas. With agricultural production for the domestic supply, through basic foodstuffs that ensure a large part of the population is fed, characterising a source of income for many Brazilian families, playing an important role in the national economy (PORTUGAL, 2004).

Veiga et al. (2001) emphasise the importance of the presence of family farming in the Brazilian countryside, given that a rural region will have a more dynamic future the greater the capacity to diversify the local economy driven by the characteristics of its agriculture. Oliveira (2000) highlighted the advantages of family production as an ideal and privileged space for consolidating sustainable agriculture.

2.3. Agroecological agriculture

Agroecology has emerged as an alternative to the problems generated by the green revolution model, based on the principles of ecological, social, economic, cultural and spatial/geographical sustainability (ALTIERI, 1998). Agroecology represents an alternative means of reproduction for family producers, based on the principles of sustainable agroecosystems and solidarity between producers.

For agroecology to really bring about a transformation in the current production moulds, it is essential that it meets the dimensions of sustainability, and it is on these dimensions that strategies can be created that will contribute to a new dynamic in rural areas (CAPORAL; COSTABEBER, 2004). In this sense, agroecology is based on two central objectives: integration and balance. integration refers to bringing together academic and scientific knowledge with that resulting from experimentation - empirical knowledge - experienced over generations by farmers.

The aim is to strike a balance between the social, environmental and economic dimensions through management techniques that combine economic development with the rational and sustainable use of available resources, enabling a balance between plants, soil and the environment.

It can be seen that agroecology aims to re-establish harmonious relations between man and his natural environment, linking what has been worked on and created by the farmer with scientific knowledge. It is the farmer who empirically knows the specificities of the agro-ecosystems in which he carries out his agricultural activities. Therefore, the great challenge is still to rescue peasant knowledge, accumulated over generations, combined with research into sustainable agriculture, which would result in a new production system with agroecological characteristics (ALTIERI; NICHOLLS, 2003).

Traditional agriculture has built up an accumulation of practices and knowledge that is of significant importance, as many farmers who have not been able to integrate into the modern production system have retained a considerable body of ecological practices, related to the technical aspects of production and the very social organisation of these farmers (CANUTO,

2003).

The agroecology proposal is complex and the interests seem contradictory at first. The challenge is to promote a transformation that goes beyond considering agroecology merely as a differentiated production strategy for family farmers. However, reality indicates that there is a long way to go for agroecology to materialise as a possibility for effective transformation in rural areas.

Strengthening sustainable family farming suggests that the old concepts of low-income farming, small-scale production and subsistence farming need to be overcome, as these have not helped to resolve the process of integrating farmers into the competitive market. In this sense, family farming should be understood more broadly as a segment that has economic and social clout (NATOI, 2001).

In Olalde's (2004) view, there is a consensus on building sustainable agriculture, seeking to value social and environmental aspects as well as economic ones. Silva and Marafon (2005) add that family farming symbolises the generation of jobs in rural areas, making it the main form of economic activity for many families, which makes it possible to contribute to food security, as well as environmental, economic and social issues.

Agroecological production as a productive conversion strategy, despite the difficulties and problems it faces, is remarkably important for the reproduction of family farming, for preserving the environment and society's quality of life. According to Salamoni and Gerardi (2001), this production system can guide the development of agriculture in a more harmonious way, based on the principles of sustainability.

2.4. Várzeas de Sousa Irrigated Perimeter

The Várzeas de Sousa Irrigated Perimeter is an initiative of the Paraíba State Government with the aim of boosting and dynamising agriculture in its area of influence, with positive effects on the state economy through actions aimed at developing agricultural and agro-industrial activities (SCI, 2012).

The Perimeter's water source is the Coremas - Mãe D'água dam complex, with a reserve capacity of 1.36 billion cubic metres of water. From the Coremas-Mãe D'água reservoir, the water is conveyed through the Adductor Canal (Canal da Redenção), which is 37 kilometres long and has a flow capacity of $4m^3/s$. As determined by the state government, the operation and maintenance of the canal is the responsibility of AESA - the state water agency (MELO et al., 2010).

The Várzeas de Souza, as they are known, are highly suitable lands for irrigation as they are made up of deep soils, flat topography, subject to high sunshine, low atmospheric humidity and where strong winds are not detected (COSTA FILHO, 2005).

Its water structure is made up of a compensation reservoir; an interconnection canal; an electrical substation; a pumping station, booster pumps, electrical control panels, a control desk with a manual and automation system, among other tools.

The project currently has 173 small producers with plots of between 5 and 10 hectares who are organised into 14 Associations founded to make it possible to raise funds for social purposes. Houses have been built on the 173 plots, with electrification and a water supply.

Training courses on commercialising production and organic farming have also been held for these farmers. In this area there are around 220 ha of crops such as bananas, coconuts, corn, beans and vegetables planted with their own resources and marketed in the region (MELO et al., 2010).

Table 1. Current division of PIVAS plots.

Current division of plots in the Várzeas de Sousa Irrigation Project (PIVAS)		
Class contemplated	**Plot size (ha)**	**Total (ha)**
16 business plots	144 ha	2.307,38
178 small producers' plots	5 ha	992,63
6 plots for the settlement of INCRA	166 ha	998,75
2 EMEPA plots	15 ha	30,00
1 batch from IFPB Sousa **Campus**	10 ha	10,00
16 plots with low suitability for irrigation	5 ha	80,00
Legal reserve	1,267.18 ha	1.267,18
Permanent preservation area	206.57 ha	206,57
Infrastructure	443, 24 ha	443, 24

area and impropieties	
Total	**6.335,74**

Source: Paraíba State Government. PIVAS Coordination.

CHAPTER 3

METHODOLOGY

3.1. Conducting Research

The methodological proposal adopted was Diniz's (1984) model, which identifies the internal and external elements that characterise agriculture. With emphasis on:

- ***S*** Social subsystem, which characterises the type of property, the owner, the physical structure of the property, land valuation and labour relations and the struggle for land;
- ***S*** Functional subsystem that analyses how land is used, farming techniques and crop rotation systems, and the intensity of farming;
- ***S*** Subsystem of production based on the analysis of the productivity of land and labour; the orientation of agriculture and the agricultural specialisation of farms.

The research universe was made up of producers in the Várzeas de Sousa Irrigated Perimeter District (PIVAS), which currently has 178 plots in operation, 173 of which are small producers. This made it possible to determine the sample, which was defined by accessibility and was therefore non-probabilistic. The sample consisted of 12 interviews among the 173 smallholder plots chosen at random, taking care not to interview farmers who were close to each other, but who were located in the various plots in the Várzeas.

The research was also based on a theoretical review of the theme of family farming and agroecology. In addition, information about the producers was collected from the Technical Assistance

Rural (ATER) of the Várzeas de Sousa Irrigated Perimeter District (DPIVAS).

3.2. Characterisation of the Study Area

The Várzeas de Sousa Irrigated Perimeter is located between parallels 6° 19' and 7$^{(o)}$ 24' S and meridians 37° 55' and 38° 46'W, with an average altitude of 225 m and is part of the Rio do Peixe sub-basin and the Piranhas river basin. The Várzeas de Sousa Irrigated Perimeter is located in the municipalities of Sousa and Aparecida, in the Sertão mesoregion of the state of Paraíba. At a distance of 420 km from the state capital - João Pessoa, the Perimeter region is connected to the latter and to the other main population centres and ports in the Northeast of the country by paved roads (CÔRREA et al., 2003).

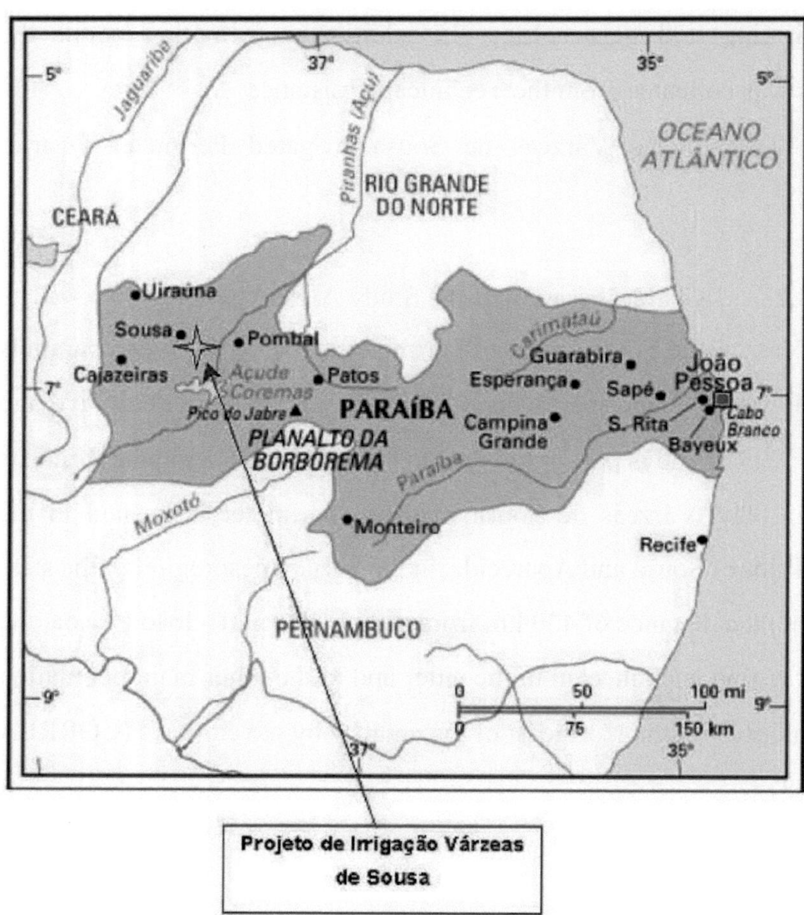

Figure 1 - Location of the Várzeas de Souza Irrigation Project. Source: <www.integracao.gov.br/.../sousa_mapa.gif>. Accessed on: 22 June 2014.

CHAPTER 4

RESULTS AND DISCUSSION

Looking at the data in Graph 1, we can see that 83 % of the family farmers interviewed were male and 17 % female, i.e. the interviews were largely conducted by men, showing that there is still a strong male predominance in access to land.

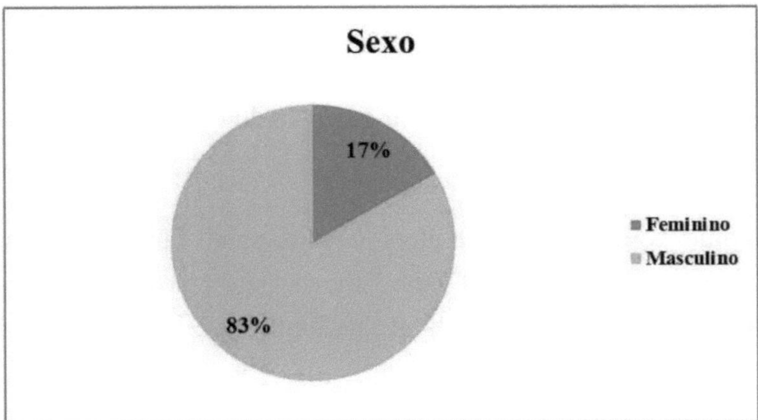

Graph 1. Sex of family farmers in the Várzeas de Sousa-PB Irrigated Perimeter.

This information confirms that traditionally there is a predominance of male land ownership in the agricultural sector. According to Brumer et al. (2008), traditionally one of the members of the family is the successor to the productive unit, with young people being the children of farmers, whether they are family farmers or not.

Graph 2 shows that 58 % of the farmers interviewed were aged between 31 and 40, 16.67 % were aged between 41 and 50 and 16.67 % were over 50. Only 8.33 % of farmers were aged between 26 and 30 and no farmers under 25 were found in the survey.

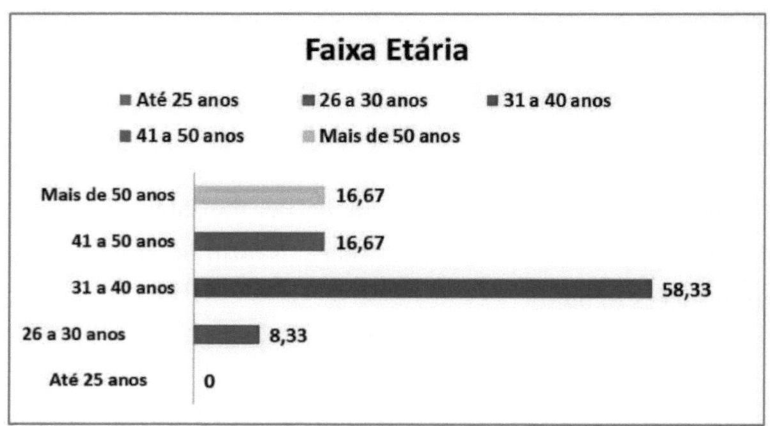

Graph 2. Age group of family farmers in the Irrigated Perimeter Sousa-PB floodplains.

When it comes to the size of the plots given to small family farmers, they range in size from 5 ha to 10 ha. Graph 3 shows that 92 % of the farmers interviewed have plots of 5 ha and 8 % of the family farmers interviewed have plots of 10 ha.

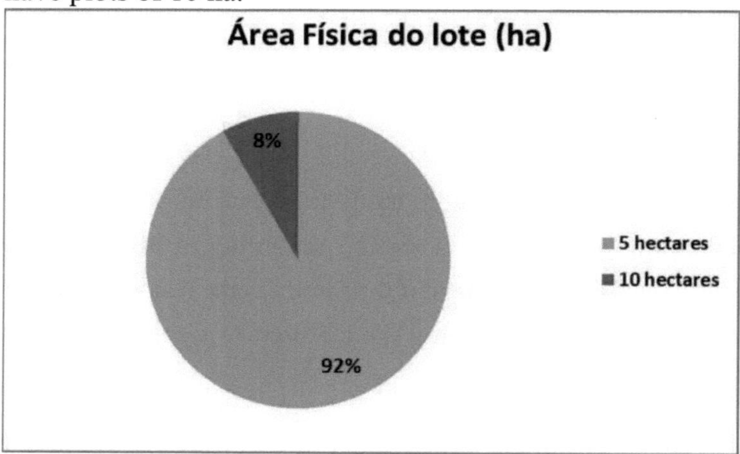

Graph 3. Physical area of the plot of family farmers in the Perimeter Irrigated

Sousa-PB floodplains.

Graph 4 shows that with regard to the labour used on the plots of family farmers in the Várzeas Irrigated Perimeter in Sousa- PB, 66 % of agricultural activities are carried out by day labourers, 17 % by salaried workers and only 17 % of agricultural activities are carried out by the families themselves.

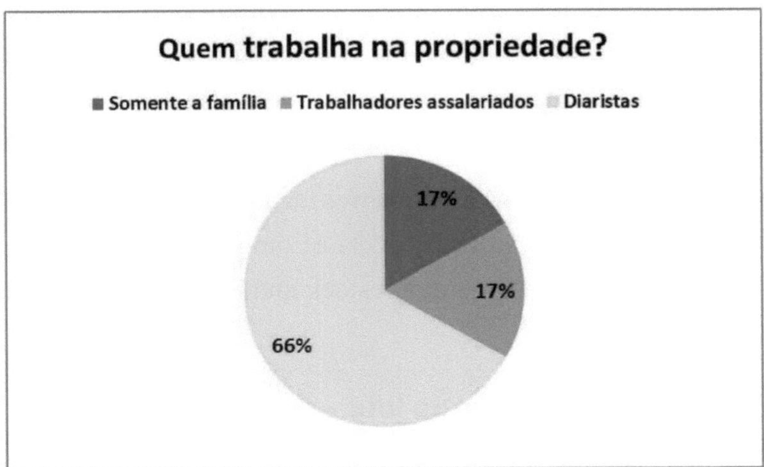

Graph 4. Labour used on family farmers' plots in the Várzeas de Sousa-PB Irrigated Perimeter.

This data confirms what Guilhoto et al. (2005) point out, that the family farming sector is always remembered for its importance in absorbing employment, as well as food production, especially for self-consumption, focussing more on social than economic functions.

According to Melo (2010), the family farm can be understood as a basic system for analysis, but one that is diverse and endowed with endogenous and exogenous relations/interactions, where the producer, his production unit and his family are the central parts of the investigation.

When asked if they had any other source of income, 50 % of the farmers said yes and 50 % said no. Similar results were found by Luz et al. Similar results were found by Luz et al. (2010) who, when carrying out a study with a community of family farmers in the municipality of Pau D'Arco - PA, all the farmers said they had an income of between one and two minimum wages, but had to supplement their income with activities outside the plot and with help from federal government assistance programmes such as Bolsa Família.

According to Sparovek (2003), income is essentially individual, with each family having a different income and labour force composition and, in the same project, there may be different production and marketing systems.

Another interesting piece of information was the farmers' use of the land on their plots, which as Graph 5 shows: 67 % of family farms grow only fruit, 25 % of plots grow both fruit and livestock and 8 % grow both fruit and vegetables.

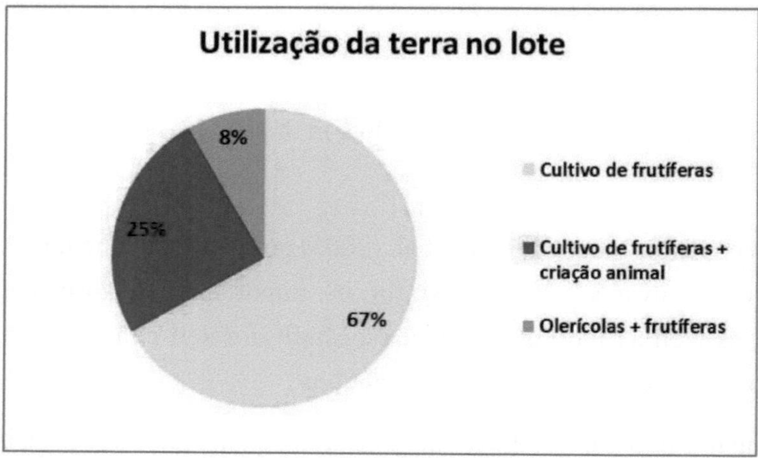

Graph 5. Land use on the plot of family farmers in the Várzeas de Sousa-PB Irrigated Perimeter.

As for the products grown by the farmers on their plots, 58.33 % of them only grow coconuts and bananas, 16.67 % coconuts, bananas and guavas, 8.33 % grow coconuts, mangoes, bananas, guavas, cashews, acerola, tomatoes, peppers, lettuce, coriander and onions, 8.33 % grow coconuts and bananas and raise sheep and 8.33 % grow coconuts and bananas and raise poultry (Table 1).

Table 1. Products grown by family farmers in the Várzeas de Sousa-PB Irrigated Perimeter.

Products grown	Percentage (%)
Coconut + banana + guava	16,67
Coconut + banana	58,33
Coconut + mango + banana + guava + cashew + acerola + tomato + chilli + lettuce + coriander + onion	8,33
Coconut + banana + sheep farming	8,33
Coconut + banana + poultry	8,33

According to Lima et al. (2001), the development of agriculture in Paraíba has been significant, especially in the Piranhas River Basin (São Gonçalo Irrigated Perimeter and Baixada de Sousa), making it a component of great economic importance for the region, given that its production is a source of income for countless families and is responsible for part of the supply of fruit to the country's major urban centres.

As family farmers who carry out a variety of agricultural and livestock activities, the interviewees were asked about their use of chemical inputs, and it was found that 75% of the interviewees replied that they do use these and only 25% replied that they do not use chemical products, as can be seen in Graph 6.

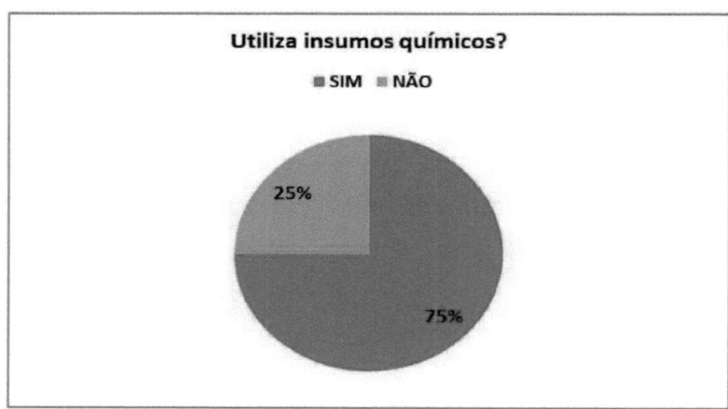

Graph 6. The use of chemical inputs by family farmers in the Várzeas de Sousa-PB Irrigated Perimeter.

Of the farmers who said they used chemical products, 89 % use both pesticides and chemical fertilisers and 11 % use only chemical fertilisers on their plots, as shown in Graph 7.

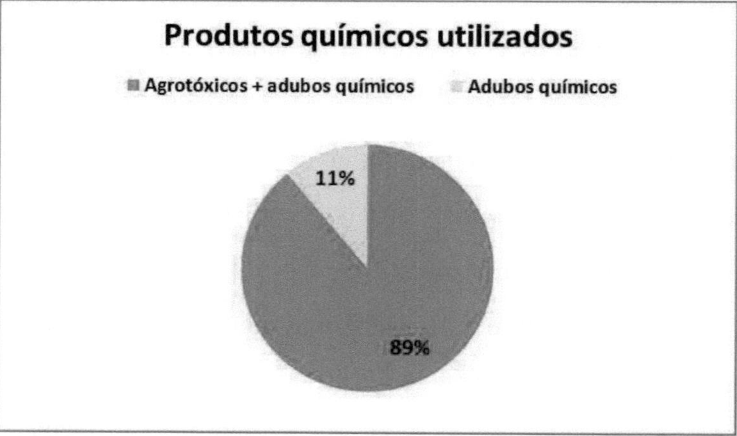

Graph 7. Which chemical products are used by family farmers in the Várzeas de Sousa-PB Irrigated Perimeter.

Brazil has been considered the world's biggest consumer of chemical inputs, according to London (2011). Research shows that between 2001 and 2008 the sale of agricultural poisons jumped from just over US$ 2 billion to over US$ 7 billion. There are approximately 366 types of chemical inputs registered for use in agriculture alone, belonging to more than 200 different groups, which give rise to 1,458 products formulated for sale on the market (PELAEZ, 2009).

During the field survey of family farmers, they were asked if they had ever heard of agroecology. 92% of those interviewed had heard of agroecology and only 8% said they had not yet heard of agroecology.

Even in the face of an apparently satisfactory result, it can be pointed out that there is a small proportion of farmers who are still unfamiliar with the term "agroecology". This result can be justified by the fact that a good proportion of our interviewees are over four decades old, since according to Gliessman (2001), it was in the early 1980s that the term "agroecology" emerged as a methodology and a distinct conceptual framework for the study of agroecosystems.

Graph 8. Of the family farmers in the Várzeas de Sousa-PB Irrigated Perimeter, which have heard of agroecology.

Casado, Sevilla-Guzmán and Molina (2000) believe in an agroecology based on the principles of sustainability, defending the idea that sustainable rural development strategies based on agroecology should occur endogenously, by strengthening mechanisms of resistance to the discourse of agrarian modernity.

Based on the concept of more sustainable agriculture, which involves the proper management of natural resources, avoiding environmental degradation so as to enable human needs to be met for current and future generations , we asked farmers if they adopted any organic practices, 92 % said yes and 8 % said no, as can be seen in Graph 9.

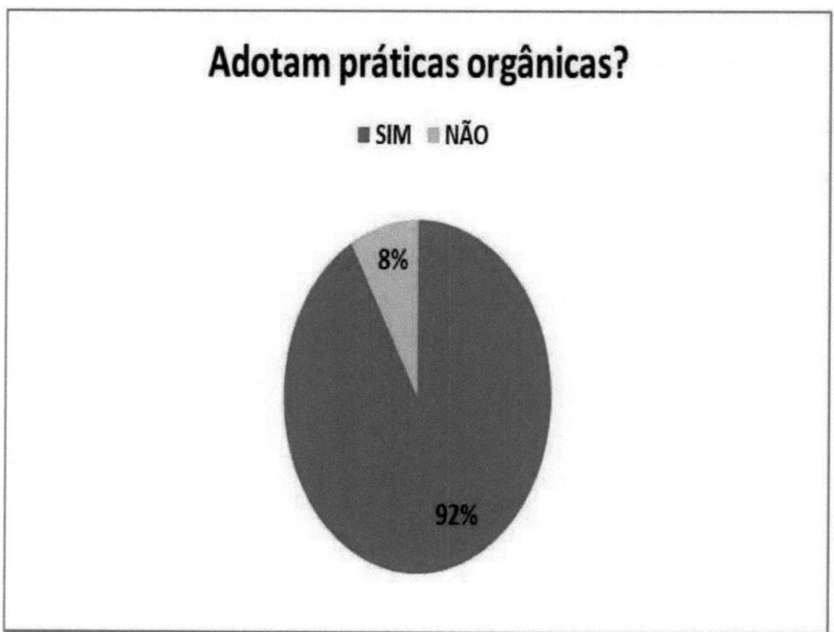

Graph 9. Adoption of organic practices on plots in the Várzeas de Sousa-PB Irrigated Perimeter.

According to Vasconcelos et al. (2013) in a study carried out in the Vista Alegre settlement in the municipality of Quixeramobim, CE, it was found that 100% of farmers use intercropping and organic fertiliser. 89 % of the settlers do not use pesticides on their crops or clear areas for planting their crops and 78 % of the families do not burn their plantation or pasture areas in the settlement.

The adoption of organic practices represents an innovative attitude in rural settlements, where family farmers establish a more balanced relationship with the environment in which they live (BRASILEIRO, 2009).

When asked which organic practices they adopted on their plots, the following results were obtained, shown in Table 2: 54.55 % of farmers only do organic fertilisation; 16.67 % use organic fertilisation and use gypsum for soil correction; 9.09 % do organic fertilisation and use biofertiliser; 9.09 % use organic fertilisation, mulching and biofertiliser; 9.09 % use organic fertiliser, compost, mulch and biofertiliser on their crops and 9.09 % of those interviewed use organic fertiliser, compost and biofertiliser on their crops.

Table 2. Organic practices adopted by family farmers in the Várzeas de Sousa-PB Irrigated Perimeter.

Organic practices adopted	Percentage (%)

Organic fertiliser	54,55
Organic fertiliser + gypsum	16,67
Organic fertiliser + biofertiliser	9,09
Organic fertiliser + mulch + biofertiliser	9,09
Organic fertiliser + compost + biofertiliser dead + top dressing	9,09
Organic fertiliser + compost + biofertiliser	9,09

According to Fontanétti et al. (2010) this set of practices, once implemented, is essential for the good performance of family farming and for maintaining and/or improving the fertility of agroecosystems. Because they are agroecologically based, they become important tools for the sustainability of agricultural production, and are aimed at maintaining or increasing the fertility of production systems and lowering the level of dependence on external inputs and reducing production costs.

With regard to the marketing practices of family farmers in the Várzeas de Sousa Irrigated Perimeter, shown in Graph 10: 58.33 % sell their produce to middlemen (traders who resell the produce in other cities in the state and/or outside the state); 16.67 % use part of the produce for consumption and sell another part to middlemen; 8.33 % also use part of the produce for consumption and sell another part at street markets; 8.33 sell all the produce at street markets and 8.33 % sell their produce directly to supermarkets in the city of Sousa and surrounding towns.

According to Cazani and Machado (2010), farmers with access to open-air markets generate a better income for their families, with direct sales adding better value to their products, contributing to the development of the municipality and preventing rural exodus.

Graph 10 - Destination of the production of family farmers in the Várzeas de Sousa-PB Irrigated Perimeter.

With regard to the use of agricultural machinery and equipment by family farmers: 50 % of those interviewed use a mechanical brushcutter, knapsack sprayer and hoe; 16.67 % use a mechanical brushcutter and knapsack sprayer; 8.33 % use a mechanical brushcutter and hoe; 8.33 % use a tractor, mechanical brushcutter, knapsack sprayer and hoe and 8.33 % use a mini tractor, atomiser, mechanical brushcutter and hoe, according to table 3.

Table 3. Equipment used by family farmers in the Várzeas de Sousa-PB Irrigated Perimeter.

Agricultural equipment used	Percentage (%)
Mini tractor + atomiser + mechanical brush cutter + hoe	8,33
Tractor + mechanical brushcutter + knapsack sprayer + hoe	8,33
Mechanical brushcutter + hoe	8,33
Mechanical brushcutter + knapsack sprayer + hoe	50,00
Tractor + mechanical brushcutter + knapsack sprayer	8,33
Mechanical brushcutter + knapsack sprayer	16,67

During the field activities, farmers were also asked if they had ever needed loans or financing, using some of the credit lines made available by governments to improve or increase production on their plots, and 50% said yes and 50% said no.

According to Bezerra and Silva (2010), one of the major problems for farmers is the lack of effective mechanisms to ensure that peasants can increase the productivity of their land and commercialise their products, and that if these problems are identified, it is necessary to build alternatives and actions that can improve the living conditions of these rural workers.

During the questionnaire, we asked whether family farmers received technical assistance, and 92 % said they did and 8 % said they did not receive technical assistance, as shown in Graph 11.

According to the Audit Report of the Court of Auditors of the State of Paraíba, the management of the Irrigated Perimeter is the responsibility of the State Secretariat for the Development of Agriculture and Fisheries - SEDAP, but technical assistance is currently carried out by the company PROJETEC, which was contracted with the aim of: promoting the administration, operation and initial maintenance of the irrigation infrastructure; organising the management entity that will be responsible for the self-management of the Perimeter; and providing technical assistance and rural extension services for small producers, with this contract running until 2015.

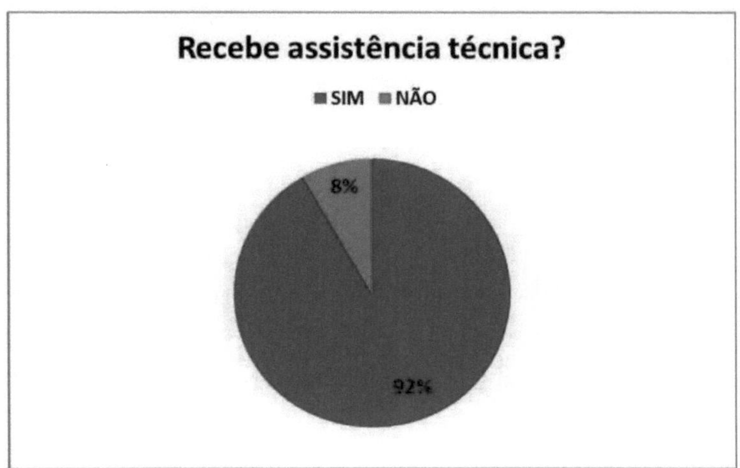

Graph 11: Producers receive technical assistance from the Várzeas de Sousa-PB Irrigated Perimeter.

According to Buainain (2007), when he carried out research with a representative sample of family farmers in different states in the Northeast, he concluded that technical assistance is one of the factors that reduces inefficiency in the use of available resources and that producers who receive monthly technical assistance have a lower degree of inefficiency in their production activities as a whole.

According to Graph 11, when family farmers were asked if they participate in an association, union, cooperative or other type of organisation, 83 % of respondents said yes and 17 % said they don't participate in any social organisation.

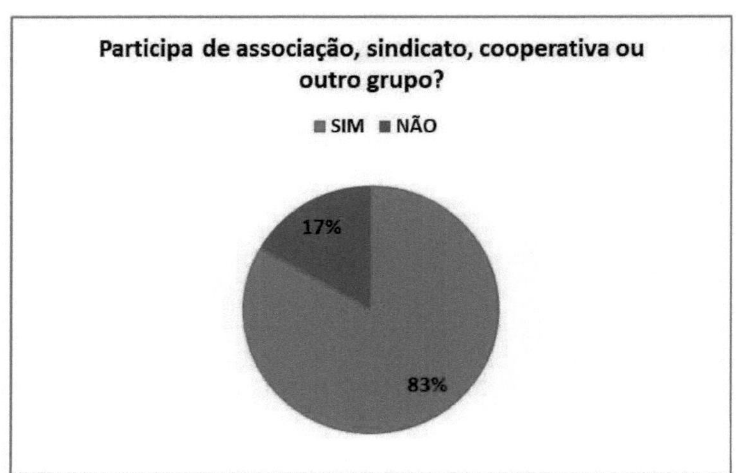

Graph 11: Participation in an association, union, cooperative or other group in the Várzeas de Sousa-PB Irrigated Perimeter.

Problems in the organisation of work, production, marketing, the acquisition and use of machinery or equipment are just a few examples of actions that can be resolved through the efforts of an organisation's own members or cooperative members.

According to Jara (1998), working within the local economic possibilities, with social stimuli (cultural, relationship, religious, among others) is essential to guarantee the sustainability of development, as well as taking responsibility for the local environmental resources, guaranteeing the well-being of future generations.

CHAPTER 5

CONCLUSION

Based on the data collected from family farmers in the Várzeas de Sousa Irrigated Project, it was found that there is a male dominance in land ownership, the majority of plots have 5 hectares, and that there is a high demand for day labourers in the irrigated perimeter, since family farming is also characterised by the absorption of employment in the countryside, thus avoiding the rural exodus.

According to the data obtained in the survey, the majority of those interviewed, 92%, adopt agroecological practices, including organic fertilisation, composting and the use and application of biofertiliser. It should also be noted that the majority of farmers only grow fruit trees, such as coconuts and bananas, and only 25% use chemical products such as poisons and chemical fertilisers.

The term agroecology is widespread among farmers, but a small minority are still unaware of it, and it is similar to the organic practices they adopt, where a large proportion of them say they use organic fertilisers.

It was found that much of the produce is sold to middlemen and that most of the equipment used by the farmers is a mechanical brushcutter, knapsack sprayer and hoe.

Technical assistance is currently provided by the company PROJETEC, whose contract with the state ends in 2015, but the results show that this service is being provided well.

In short, the Várzeas de Sousa present satisfactory conditions for the development of agroecological practices, as it is a perimeter with fertile land, with water available for cultivation, and because the initial project was created with the aim of organic farming being the main means of agricultural development, the plots were cultivated organically.

CHAPTER 6

REFERENCES

AGUIAR-MENEZES, E. L. de; SANTOS, C. M. S.; RESENDE, A. L. S.; SOUZA, S. A. S.; COSTA, J. R.; RICCI, M. S. F. Susceptibility of coffee cultivars to insect pests and diseases in an organic system with and without afforestation. Seropédica: Embrapa Agrobiologia, 2007 (Research and Development Bulletin/Embrapa Agrobiologia).

ALTIERI, M. Agroecologia: a dinâmica produtiva da agricultura sustentável. Porto Alegre: Ed. Universidade/UFRGS, 1998. 110p.

ALTIERI, M.; NICHOLLS, C. Agroecologia: teoría y práctica para una agricultura sustentable. Mexico: UNEP and the Environmental Training Network for Latin America and the Caribbean, 2000.

ALTIERI, M.; NICHOLLS, C. I. Agroecology: Rescuing organic agriculture from an industrial model of production and distribution. Ciência e Ambiente, Santa Maria, v.1, n. 1(jul. 1990), p.141-152, 2003.

ARAÚJO, D. L.; BRESSAN, A. A.; BERTUCCI, L. A.; LAMOUNIER, W. M. The market risk of Brazilian agribusiness: a comparative analysis between the CAPM and GARCH-M models. Gestão.Org, v. 2, n. 3, p. 207-220, Sep./Dec. 2004.

BARROS, J. R. M. de; GOLDENSTEIN, L. Avaliação do processo de

Brazilian industrial restructuring. Revista de Economia Política. São Paulo, n. 2, v. 17, Apr./Jun.1997.

BARROS, A. L. M. de. Brazilian agribusiness: characteristics and challenges. Biotechnology of Reproduction in Bovines (II International Symposium on Applied Animal Reproduction). Londrina-PR, 2006.

BEZERRA, C; SILVA, R. Analysis of agrarian reform settlements under the jurisdiction of INCRA in Alagoas. Presentation to the VIII Latin American Congress of Rural Sociology, Proto de Galinhas, 2010.

BORGES, M. A percepção do agricultor familiar sobre o solo e a agroecologia. 2000. 245f. Dissertation (Master's in Agricultural Engineering) - Institute of Economics, State University of Campinas, Campinas, 2000.

BORGES FILHO, E. L. From the reduction of agricultural inputs to agroecology: the trajectory of research into more ecological agricultural practices at EMBRAPA. 2005. 279f. Thesis.

BRAZIL. Agribusiness projections: Brazil 2012/2013 to 2022/2023. Ministry of Agriculture, Livestock and Supply. Strategic Management Office. - Brasília: Mapa/ACS, 2013.

BRASILEIRO, R.S. Alternatives for sustainable development in the northeastern semi-arid region: from degradation to conservation. Scientia

Plena, v.5, n.5, 1-12p, 2009.

BRUMER, A.; PANDOLFO, G. C.; CORADINI, L. Gênero e agricultura familiar: projetos de jovens filhos de agricultores familiares na região Sul do Brasil. 2008. Available at: <http://www.fazendogenero.ufsc.br/8

/sts/ST3/Brumer-Pandolfo-Coradini_03.pdf>. Accessed on: 10 July 2014.

CAPORAL, F. R.; COSTABEBER, J. A. Agroecology: Some concepts and principles. Brasília: MDA/SAF/DATER-IICA, 2004. 24p.

CANUTO, J. C. Research and the Challenges of Agroecological Transition. Science & Environment. Santa Maria. Vol. 1, n. 1 (Jul. 1990), p. 133-140, 2003.

CASADO, G. G.; SEVILLA-GUZMÁN, E. and MOLINA, M. G. Introducción a la agroecología como desarrollo rural sostenible. Madrid: Mundi-Prensa, 2000.

CAZANI, A. L.; MACHADO, J. G. de C. F. Analysis of free markets in Tupã - SP based on the behaviour of FV consumer. In. CONGRESS OF THE BRAZILIAN SOCIETY OF RURAL ECONOMY, ADMINISTRATION AND SOCIOLOGY - SOBER, 48TH, 2010. Campo Grande - MS. Proceedings... Campo Grande - MS, 2010.

CHONCHOL, J. Food sovereignty. Advanced Studies. São Paulo: USP, v.

19, n. 55, p. 33-48, 2005.

CORRÊA, M.M.; KER, J.C.; MENDONÇA, E.S.; RUIZ, H.A. BASTOS, R.S. Physical, chemical and mineralogical attributes of soils from the Várzeas de Sousa region (PB). Revista Brasileira de Ciência do Solo, vol.27 n° 2, Viçosa, Mar./Apr. 2003.

COSTA FILHO, J. F. Evaluation of energy balance components and evapotranspiration in a semi-arid region using Landsat 5 - TM and Terra MODIS orbital images. 159p. Thesis (Doctorate in Meteorology), Federal University of Campina Grande, Campina Grande, 2005.

DINIZ, J. A. F. Geografia da Agricultura. São Paulo: Difel, 1984. 278p.

DUARTE, C. Family farming and sustainable development models. 2012. Available at: http: <//www.terraquente.net/>. Accessed on: 30 June 2014.

ESCÓSSIA, C. What is agribusiness? 2009. Available at: http: <//www.carlosescossia.com>. Accessed on: 19 Jun. 2014.

FEIDEN, A. Agroecology: introduction and concepts. In: AQUINO, A. M. de; ASSIS, R. L. Agroecologia: princípios e técnicas para uma agricultura orgânica sustentável. Brasília: Embrapa Informação Tecnológica, 2005. p.49-70.

FONTANÉTTI, A.; SANTOS, I. C. dos. Fertility management of the agroecosystem and the sustainability of family farming. Informe Agropecuário, Belo Horizonte, v. 31, n. 254, p.7-13, 2010.

FRANCO, H. Conventional Agriculture. 2010. Available at: <estudosimples.blogspot.com/p/agricultura-convencional.html>. Accessed on: 10 June 2014.

GERARDI, L. H. O.; SALAMONI, G. Para entender o campesinato: a contribuição de A. V. Chayanov. Geografia, Rio Claro, v. 19, n.2,p.123-140, 1994.

GLIESSMAN, S. R. Agroecologia: processos ecológicos em agricultura sustentável (2001) ed. Porto Alegre: Ed. Universidade/UFRGS, 653 p.

GLIESSMAN, S. R. Agroecology: ecological processes in sustainable agriculture. 3. ed. Porto Alegre: UFRGS, 2005.

GUILHOTO, J. J. M.; SILVEIRA, F. G.; AZZONI, C. R.; ICHIHARA, S. M. "The GDP of Family Agribusiness in Rio Grande do Sul". Proceedings of the XLIII Congress of the Brazilian Society of Rural Economics and Sociology. Ribeirão Preto, São Paulo, 24 to 27 July 2005.

IAMAMOTO, A.V.T. Agroecology and rural development. 2005. 79f. Dissertation (Master's Degree in Forest Resources) - Escola Superior de Agricultura "Luiz de Queiroz", Universidade de São Paulo, Piracicaba, 2005.

JARA, C. J. The sustainability of local development: Challenges of a process under construction. Brasília: Inter-American Institute for Cooperation on Agriculture (IICA): Recife: Planning Secretariat of the State of Pernambuco - Seplan, 1998.

LIMA, C.O.; BARBOSA, M.P.; LIMA, V.L.A.; SILVA, M.J. Use of TM/Landsat-5 images and thermometry in the identification and mapping of

soils affected by salts in the region of Sousa-PB. Revista Brasileira de Engenharia Agrícola e Ambiental, v.5, n.2, p.361-363, 2001.

LOPES, P. R.; LOPES, K. C. S. A. Ecologically-based production systems - the search for sustainable rural development. REDD - Revista Espaço de Diálogo e Desconexão, Araraquara, v. 4, n. 1, jul/dez. 2011.

LUZ, C. A. da.; BRITO, I. Sousa.; SILVA, M. da Cruz.; MARRAFON, A. M. de Almeida. Characterisation of family farmers in the Boa Sorte community in the municipality of Pau D'arco in south-eastern Pará through participatory diagnosis. Federal Institute of Education, Science and Technology of Pará. Conceição do Araguaia-PA Sector, 2010.

MEIRELLES, L. Food sovereignty, agroecology and local markets. Revista Agriculturas: experiências em agroecologia: AS-PTA - Assessoria e Serviços a Projetos em Agricultura Alternativa, v. 1, n. 0, p. 11-14, set.

2004.

MELO, S. T. de S., RIBEIRO, F. A., ARAÚJO, C. F., MOREIRA, E. The struggle for land and water in the floodplains of Sousa. XVI National Meeting of Geographers, Porto Alegre - RS, 2010.

MESQUITA, L.A. P. de.; MENDES, E. de P. P. Family Farming, Labour and Strategies: Women's Participation in Socioeconomic and Cultural Reproduction. Espaço em Revista. vol. 14 n° 1 jan/jun.2012 p. 14 - 23.

MOTTA, M.; ZARTH, P. Introduction. In: MOTTA, M; ZARTH, P. (org.) Formas de resistência camponesa: visibilidade e diversidade de conflitos ao longo da história. São Paulo: Ed. Unesp/Brasília: MDA/NEAD, 2008 (Coleção História Social do Campesinato no Brasil).

OLALDE, A. R. Family farming and sustainable development. 2004. Available at: <www.ceplac.gov.b>. Accessed on: 19 June 2014.

OLIVEIRA, J. T. A. Lógicas Produtivas e Impactos Ambientais: Estudo Comparativo de Sistemas de Produção. 2000. 284p. Thesis (Doctorate in Agricultural Engineering) - Unicamp, Campinas, 2000.

OLIVEIRA, A. U. The Lula government bids farewell to agrarian reform. Brasil de Fato, São Paulo, 22 Dec 2008. Available at: <http://www.brasildefato.com.br/node/3444>. Accessed on: 22 Jun. 2014.

OTANI, M. N. Caracterização e Estudo da Agricultura Familiar: o caso dos produtores de leite do município de Lagoinha, Estado de São Paulo. Informações Econômicas, São Paulo: v.31, n.4, Apr. 2001.

PELAEZ, V.; TERRA, F. H. B; SILVA, L. R. The regulation of pesticides in Brazil: between market power and the defence of health and the environment. Paper presented at the XIV Encontro Nacional de Economia Política / Sociedade Brasileira de Economia Política - São Paulo/SP, from 09/06/2009 to 12/06/2009. 22 p. Available at: <http://www.sep.org.br/artigo/1521_b91605d431331313c8d7e1098bb1dd 3 4.pdf>. Accessed on: 10 July 2014.

PORTUGAL, D. A. The challenge of family farming. 2004. Available at <http://www.embrapa.br/imprensa/artigos/2002/artigo.2004-12-07.2590963189/>. Accessed on 22 June 2014.

Várzeas de Souza Irrigation Project. Available at: <www.integracao.gov.br/.../sousa_mapa.gif >. Accessed on: 22 June 2014.

RIBEIRO, V. S.; SALAMONI, G.; COSTA, A. J. V. da. Characterisation of agroecological family farmers in the municipality of Pelotas, RS. 5th Meeting of Research Groups - Agriculture, Regional Development and Socio-Spatial Transformations. UFSM - GPET, November/2009.

SALAMONI, G.; GERARDI, L. H. de O. Principles of eco-development and their relationship with family farming. In: Teoria, técnica, espaços e atividades: temas da geografia contemporânea. Rio Claro/SP. Postgraduate Programme in Geography - UNESP; Association of Theoretical Geography. AGETEO, 2001, p.73-96.

SCHNEIDER, S.; PLEIN, C. Family farming and commercialisation. In: CASTILHO, Mara Lucy; RAMOS, José Maria (editors). Agribusiness and sustainable development. Francisco Beltrão: Calgan, 2003.

SCHROETTER, M. R. Work plan for the technicians of the commercialisation bases for family farming products and the solidarity economy (bsc's). Santa Rosa. 2010.

SCI - Secretariat for Institutional Communication of the Government of the State of Paraíba. Report. Available at <http://www.paraiba.pb.gov.br/agropecuaria-e-pesca/programas-e-acoes>. Accessed: 22 June 2014.

SIMON, Á. Engineer emphasises the importance of family farming for the economy. Mundo rural, 2014. Available at: http:<//diariocatarinense.clicrbs.com.br/>. Accessed on: 19 June 2014.

SILVA, E. S. O. da.; MARAFON, G J. Family farming in the state of Rio de Janeiro. I International Seminar on Regional Development Postgraduate Programme in Regional Development Master's Degree and Doctorate Santa Cruz do Sul-RS, 2005.

SPAROVEK, G. A qualidade dos assentamentos da reforma agraria brasileira. Editora e Gráfica Paginas & Letra, São Paulo, 2003. ISBN 88586508-26-8.

TEIXEIRA, J. C. Modernisation of agriculture in Brazil: economic, social and environmental impacts. Revista Eletrônica da Associação dos Geógrafos Brasileiros, v.2, n.2, p.21-42, Sep. 2005.

COURT OF AUDITORS OF THE STATE OF PARAIBA. Operational Audit Report on the Sousa Floodplains. Process TC n° 04338/13. Act of appointment: Ordinance No. 136 of 26 November 2012.

VASCONCELOS, J. M. G; MOURÃO, A. E. B; CAVALCANTE, A. C, R; FRANCO, F. S. Agroecological practices for living in the semi-arid region

adopted by family farmers in the sertão of Ceará. I Brazilian Symposium on Natural Resources of the Semi-Arid. Iguatu - CE, 22 to 24 May 2013.

VEIGA, J. E.; FAVARETO, A.; AZEVEDO, C.M.A.; BITTENCOURT, G.; VECCHIATTI, K. ; MAGALHÃES, R.; JORGE, R. O Brasil rural precisa de uma estratégia de desenvolvimento, Brasília: Convênio FIPE-IICA (MDA/CNDRS/NEAD), 2001. 108 p. Available at: <http://www.nead.org.br/index.php?acao=bibliotecaepublicacaoID=112>. Accessed on: 22 June 2014.

VEIGA, J. E. da. Agriculture in the modern world: diagnosis and prospects.

In: TRIGUEIRO, A (Org.) Meio ambiente no século 21: 21 especialistas falam da questão ambiental nas suas áreas de conhecimento. Rio de Janeiro: Sextante, 2003. p.199-213.

WANDERLEY, M. N. B. Historical roots of the Brazilian peasantry. In: CARNEIRO, M. J. MALUF, R. S. (org.) Para além da produção: multifuncionalidade e agricultura familiar. Rio de Janeiro: MAUAD, 2003.

Printed by Books on Demand GmbH, Norderstedt / Germany